10/12
23.00

The Human Touch

Michael Frayn was born in 1933 and began his career as a journalist on the *Guardian* and the *Observer*. His novels include *Towards the End of Morning*, *The Trick of It* and *A Landing on the Sun*. *Headlong* (1999) was shortlisted for the Booker Prize, while his most recent novel, *Spies*, won the Whitbread Novel Award. His fifteen plays range from *Noises Off* to *Copenhagen*, and most recently *Democracy*.

He is married to the biographer and critic Claire Tomalin.

Further praise for *The Human Touch*

'Learned and acute . . . Framed with all the wit and precision of his other writings.' Simon Blackburn, *Times Literary Supplement*

'With lashings of wit to leaven his erudirion, a host of funny anecdotes and some exuberantly Fraynian digressions, *The Human Touch* develops its vision of a world spun from human stories.' Boyd Tonkin, *Independent*

'Compelling reading . . . . Frayn is a disarmingly persuasive guide as we accompany him on an exhilarating intellectual journey.' Lisa Jardine, *The Times*

'Gives hope to the layman, among many delights . . . that the gap between the "two cultures" of humanities and science is still bridgeable.' Tim Adams, *Observer*

'A work of popular erudition . . . This book forces the mind into all sorts of contortions, not all of them comfortable. But, my goodness, it's worth it.' William Leith, *The First Post*

'Relaxed but lucid, profuse but engrossing. You could not hope for a more elegant foray into the fundamental questions of philosophy'. Joanthan Rée, *Evening Standard*

D0018805